ANIMAL ECO INFLUENCERS

GRAY WOLVES

Yellowstone's Hunters

MEGAN BORGERT-SPANIOL

Consulting Editor, Diane Craig, M.A./Reading Specialist

An Imprint of Abdo Publishing
abdobooks.com

ABDOBOOKS.COM

Published by Abdo Publishing, a division of ABDO, PO Box 398166, Minneapolis, Minnesota 55439.

Printed in the United States of America, North Mankato, Minnesota
102019
012020

Design: Kelly Doudna, Mighty Media, Inc.
Production: Mighty Media, Inc.
Editor: Liz Salzmann
Cover Photographs: Shutterstock Images
Interior Photographs: Francois Gohier/ARDEA p. 11; Getty Images/iStockphoto, pp. 4, 5, 6, 7, 10, 11, 14; Shutterstock Images, pp. 3, 4, 5, 6, 7, 8, 9, 10, 11, 12, 13, 14, 15, 16, 17, 18, 19, 20, 21, 22, 23

Publisher's Cataloging-in-Publication Data
Names: Borgert-Spaniol, Megan, author.
Title: Gray wolves: Yellowstone's Hunters / by Megan Borgert-Spaniol
Other title: Yellowstone's hunters
Description: Minneapolis, Minnesota : Abdo Publishing, 2020 | Series: Animal eco influencers
Identifiers: ISBN 9781532191862 (lib. bdg.) | ISBN 9781532178597 (ebook)
Subjects: LCSH: Gray wolf--Juvenile literature. | Yellowstone National Park--Juvenile literature. | Carnivora--Behavior--Yellowstone National Park Region--Juvenile literature. | Food chains (Ecology)--Juvenile literature. | Predatory animals--Behavior--Juvenile literature. | Animal ecology--Juvenile literature. | Wildlife habitats--Juvenile literature.
Classification: DDC 599.773--dc23

CONTENTS

ECO INFLUENCERS

What is an eco **influencer**? It is an animal that can change its **ecosystem**. All members of an ecosystem affect one another.

Prairie dogs dig tunnels in their prairie ecosystem. This creates healthy soil.

Sea otters help shape their coastal ecosystem. They maintain **kelp** forests by eating sea urchins.

Gray wolves are eco **influencers**. They prey on elk in Yellowstone National Park. This means elk eat fewer plants. So, there are more plants for other wildlife to eat. Gray wolves are Yellowstone's hunters!

THINK!

Can you think of other animals that help shape their **ecosystems**?

PARK PROTECTION

Between 80 and 100 gray wolves live in Yellowstone. There, they are protected from hunters. Scientists study the gray wolves in Yellowstone to learn more about wolf behavior.

Wolves also live in many other areas across the **Northern Hemisphere**.

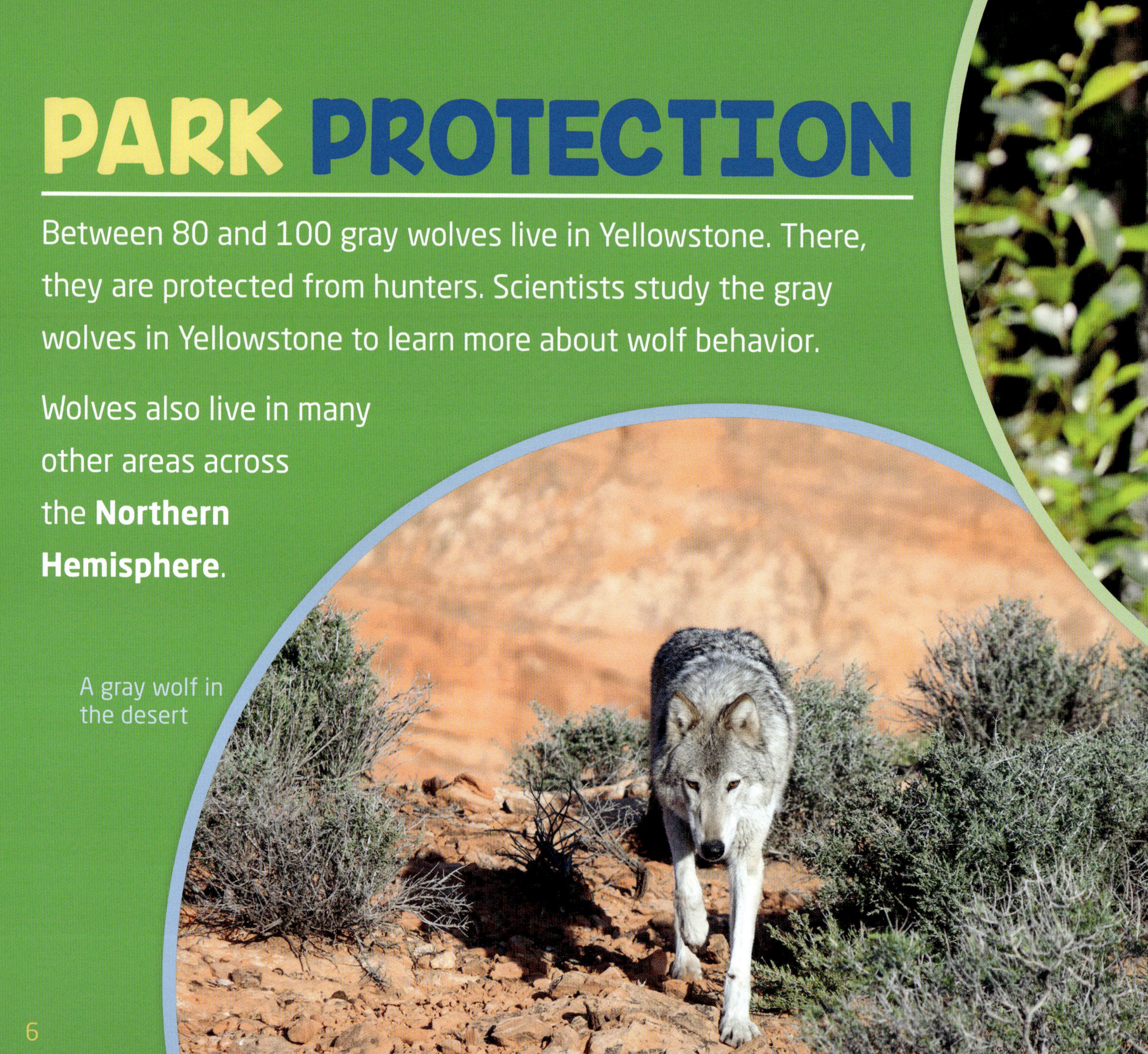

A gray wolf in the desert

GRAY WOLF HOMES

Gray wolves can survive in many different **habitats**.

Deserts

Forests

Grasslands

Mountains

Tundra

Gray wolves in the forest

WHERE GRAY WOLVES LIVE

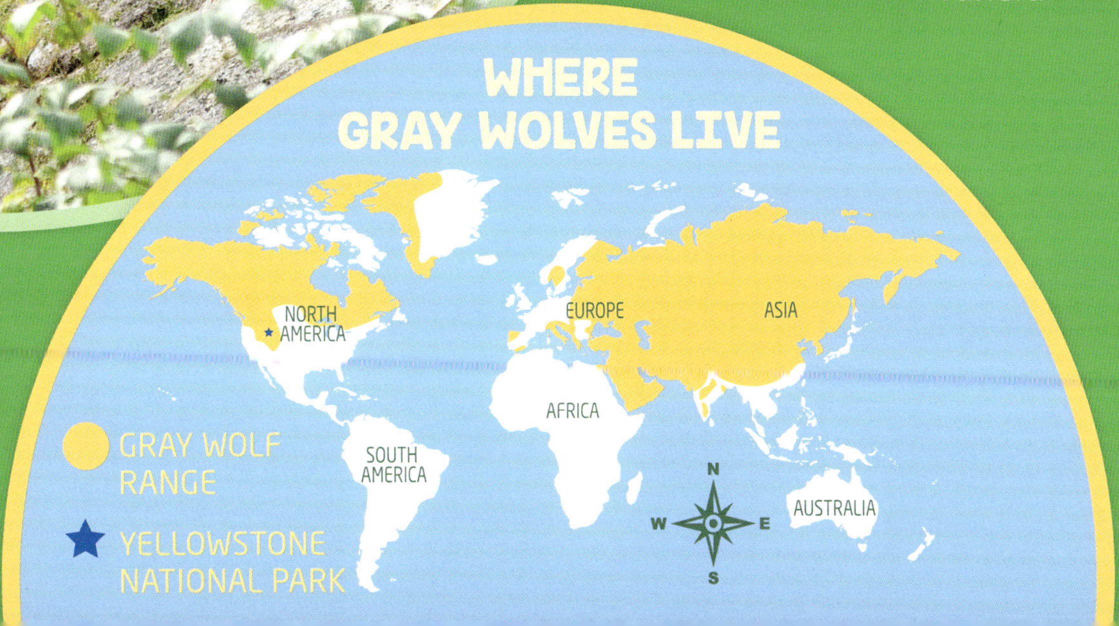

HUNT AND SCAVENGE

Gray wolves need plenty of space. They are built to move! Gray wolves have long legs that help them **trot** for hours at a time. These **carnivores** stay on the move to find prey.

Elk

As a pack, gray wolves hunt large animals. These include elk and deer.

Wolves hunting alone or in pairs hunt smaller prey. These include rabbits and beavers.

Gray wolves are also **scavengers**. They eat animals that have died of other causes.

Rabbit

Wolf scavenging

THINK!
What plants or animals do you eat? Where does your food come from?

APEX PREDATORS

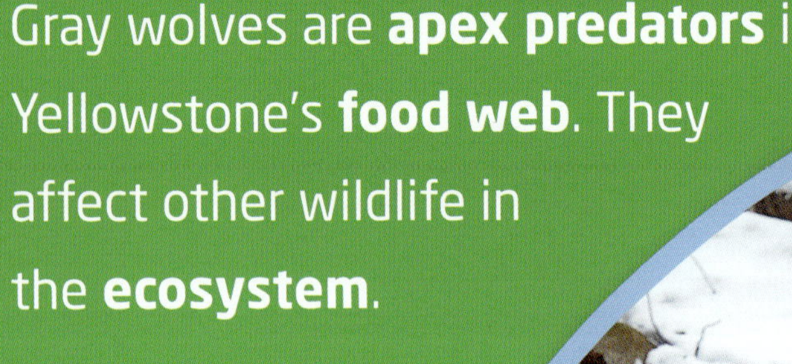

Gray wolves are **apex predators** in Yellowstone's **food web**. They affect other wildlife in the **ecosystem**.

Prey for wolves in Yellowstone include bison.

A fox in Yellowstone

Ravens eating a dead elk

HOW DO GRAY WOLVES AFFECT THE FOOD WEB?

They choose weak and sick prey.

This can help stop the spread of illness. It also means strong and healthy animals survive to have young.

They kill coyotes.

This means there are fewer coyotes to hunt animals such as rabbits and mice. In turn, these animals **attract** foxes, hawks, and other predators.

Their kills feed other **scavengers**.

Gray wolves eat until they are full. The leftovers help feed birds, bears, and other scavengers.

BEAUTIFUL BANKS

Wolves can also **influence** the land. It starts with elk!

Elk eat trees and bushes that grow near rivers and streams. The elk could wipe out the plants along the banks.

This changes when wolves are around. To avoid wolves, elk move more. They spend less time eating along rivers and streams. This allows more plants to grow on the banks.

THINK!
How do your daily activities affect the land around you?

BIRDS, BEAVERS, AND BEYOND

Renewed plant growth along banks leads to even more changes.

The trees and flowers **attract** birds and insects.

Dragonfly

Mountain bluebird

Beaver lodge

River otter

New tree growth also supports beavers. Beavers eat wood. They also use wood to build dams and lodges.

Beaver dams create **wetlands**. The wetlands support fish, otters, and other **aquatic** animals!

WOLVES AND RIVERS

Plant growth can also affect how rivers and streams flow. Plants help reduce **erosion** along banks. This keeps rivers and streams from changing size and course.

Without plants, erosion causes rivers to widen. This makes them shallower and warmer.

Rivers with plants are deeper. And shade from plants keeps the water cool.

Warm, shallow rivers cannot support much aquatic life.

Trout and other aquatic animals live in deep, cool rivers.

RIPPLE EFFECT

Scientists study the wolf population in Yellowstone. They have learned that wolves do more than kill prey. Wolves create an **ecosystem** that supports a lot of wildlife! This happens in a **ripple effect**.

Reduced **erosion** and more plants to provide shade

Rivers and streams stay deeper and cooler.

Rivers, streams, and wetlands **attract** more **aquatic** life.

Beavers cut down trees to build dams.

Beaver dams create **wetlands**.

THINK!

Draw a picture of a river where wolves do not live. Then draw the river after wolves start living there.

WOLF CONSERVATION

Conservation programs have helped save gray wolves. Wolves are protected in parks such as Yellowstone. Outside of these areas, humans are the greatest danger to gray wolves. But **experts** know how important gray wolves are to their **ecosystem**. These people keep working to protect gray wolves!

Farmers worry that wolves will kill sheep and other livestock.

Gray wolves in Yellowstone

ECO INFLUENCER FACT SHEET

Common name: Gray wolf

Class: Mammal

Life span in the wild: 6 to 8 years

Population trend: Stable

Diet: Carnivore

Size in relation to humans:

FUN FACTS

A gray wolf's sense of smell is 100 times stronger than a human's.

Gray wolves live in groups called packs. About ten wolves live in a pack.

A gray wolf can eat more than 20 pounds (9 kg) of meat in a single meal.

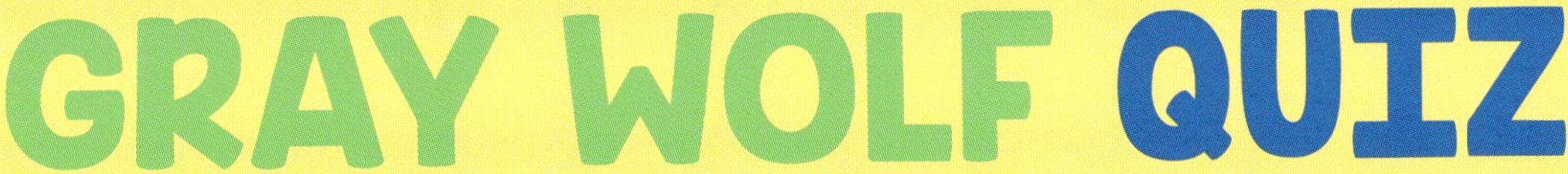

1. What feature helps a gray wolf **trot** for a long time?

   A. thick body

   B. long legs

   C. bushy tail

2. Plant growth helps reduce **erosion** along banks.

   True or **false**?

3. What is the greatest danger to gray wolves?

Answers: 1. B 2. True 3. Humans

GLOSSARY

apex predator—a predator that is not hunted by any other animals.

aquatic—living in water.

attract—to cause someone or something to come near.

carnivore—an animal that eats mainly meat.

conservation—the act or process of saving and protecting something.

ecosystem—a group of plants and animals that live together in nature and depend on each other to survive.

erosion—wearing away of the land often caused by water or wind.

expert—a person very knowledgeable about a certain subject.

food web—the feeding relationships between different organisms in a community.

habitat—the area or environment where a person or animal usually lives.

influence—to cause something to change.

kelp—a large, brown seaweed.

Northern Hemisphere—the half of Earth that is north of the equator.

ripple effect—a situation in which something causes a series of changes.

scavenger—one that feeds on whatever garbage or dead animals it finds.

trot—to run at a moderately fast pace.

wetland—a low, wet area of land such as a swamp or a marsh.